LIGHT
AND
SOUND

Chris Oxlade

HODDER
Wayland

An imprint of Hodder Children's Books

Titles in the *Science Files* series:

Electricity and Magnetism • Forces and Motion • Light and Sound • The Solar System

Science Files is a simplified and updated version of Hodder Wayland's
Science Fact Files.

For more information on this series and other Hodder Wayland titles,
go to www.hodderwayland.co.uk

Text copyright © Hodder Wayland 2005

Editor: Katie Sergeant
Designer: Simon Borrough
Typesetter: Victoria Webb
Illustrator: Alex Pang

First published in Great Britain in 1999 by Macdonald Young Books,
an imprint of Wayland Publishers Ltd
This edition updated and published in 2005 by Hodder Wayland,
an imprint of Hodder Children's Books

Oxlade, Chris
   Light and sound. – (Science files)
   1.Light – Juvenile literature 2.Sound – Juvenile literature
   I.Title
   534

ISBN 0750247118

Printed in China by WKT Company Ltd

Hodder Children's Books
A division of Hodder Headline Limited
338 Euston Road, London NW1 3BH

Cover picture: Prisms.
Endpaper picture: A dazzling firework display.
Title page picture: The artificial beam from a lighthouse is outshone by natural lighting.

We are grateful to the following for permission to reproduce photographs: Digital Stock 36 bottom (Marty Synderman); Digital Vision 9, 15, 34; Getty Images/Photodisc Green cover (Royalty-free/ Lawrence Lawry); Science Photo Library 8 left (Alex Bartel), 8 bottom (Martin Bond), 11 bottom (Keith Kent), 12 bottom (G Antonio Milani), 14 (Cortier/BSIP), 16 (Jerome Wexler), 18 top (Alfred Pasieka), 20 (Claude Nuridsany & Marie Perennou), 21 (Omikron), 22 (David Scharf), 26 (Robert Holmgren/Peter Arnold Inc), 27 (Phillippe Plailly/Eurelios), 28 top and bottom left (Phillippe Plailly), 28 bottom right (David Parker), 30 (Will & Deni McIntyre), 32 (Volker Steger), 33 (US Air Force), 36 top (Carlos Munoz-Yague/Eurelios), 37 (Dr Morley Read), 38 Prof P Motta/La Sapienza). 39 both (Prof C Ferlaud/CNRI), 42 (Dr Jeremy Burgess), 43 bottom (CS Langlois/Publiphoto Diffusion); The Stock Market 11 top (John Martin), 12 top (Sanford/Agliolo), 18 bottom, 35 top (Milt & Patti Putnam), 43 top (Torlief Svensson); Tony Stone 25 (Robert E Daemmrich), 31 (Tom Raymond), 35 bottom (Demetrio Carrasco).

Contents

Words in **bold** can be found in the glossary on page 44.

Introduction

The world would be a very different place without light and sound. Light and sound let us see and hear the world and what is happening in it. We use our senses of sight and hearing to find our way, to communicate, to avoid danger, and to enjoy our surroundings. So do many animals.

Humans have learned how to use light and sound for a wide range of scientific, industrial and medical applications. For example, we use **laser beams** to cut metals, and we use sound to see unborn babies inside their mothers.

HISTORY FILE

THE NATURE OF LIGHT
Long ago people didn't know about light waves or light energy. They thought we could see things because they were lit up by light coming from our eyes. Now we know that light comes from sources such as the Sun and bounces off objects into our eyes. The first scientist to realise this was called Alhazen. He lived in Persia a thousand years ago.

ENERGY, LIGHT AND SOUND

Energy makes things happen. It makes things move, it makes machines work, and it makes substances change. Nothing can happen without energy. Light and sound are both forms of energy. They carry energy from place to place in the form of waves.

A dazzling light display made by fireworks.

This machine captures the energy in sea waves to make electricity.

FACT FILE

ALL ABOUT WAVES

• Waves spread out in every direction from their source (where they are made) unless something gets in the way.

• Waves of the same kind always travel at the same speed in the same substance. For example, sound waves always travel at 340 metres per second in air.

• Waves normally change speed and direction when they travel from one substance to another.

• The **amplitude** of a wave is the height of a peak or the depth of a trough.

• The **wavelength** of a wave is the distance from one wave peak to the next wave peak, or one wave trough to the next trough.

• The **frequency** of a wave is the number of waves that pass by every second. It is measured in hertz (Hz).

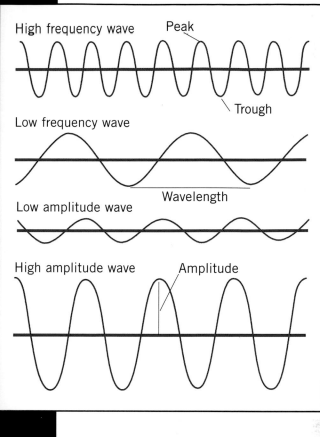

Waves of different wavelengths, frequencies and amplitudes.

WAVE ENERGY

We are used to seeing waves on water. Each wave has a **peak** like a hill and **trough** like a valley. If you watch the water carefully, you'll see that the water does not move along in the direction of the waves, but only up and down as the waves pass by. So the waves **displace** the water up and down. They carry energy through the water.

Light and sound are also forms of wave energy, but we can't see them as they travel. And they are very different from each other. Sound waves can only travel through substances, but light waves can travel through a **vacuum** (a vacuum is a space where there is nothing at all, not even air). Light is a million times faster than sound.

You couldn't hear this spacecraft because sound can't travel in a vacuum.

What is Light?

Light is easy to see, but difficult to describe! It is a type of **electromagnetic radiation.** All electromagnetic radiation is made up of waves of electricity and waves of magnetism travelling together. Other forms of electromagnetic radiation are radio waves, X-rays, microwaves and infrared (heat) rays. Together these waves or rays make up a whole family of electromagnetic radiation, which is called the electromagnetic spectrum. They all travel at the same speed, but they have different wavelengths.

SOURCES OF LIGHT

Most of the light we see comes from the Sun. Without sunlight life on Earth would be impossible. Light normally comes from objects that are so hot that they give out energy in the form of light. Some light bulbs have a wire called a filament inside that glows when electricity flows through it, and candles have a hot, glowing flame.

Core

Heat spreads from core as radiation

Heat moves to surface in swirling gas

The Sun is a giant ball of glowing gas. Huge amounts of energy come from *nuclear reactions* in the Sun's centre. Most of this energy leaves the Sun as light and heat.

FACT FILE

POLARISED LIGHT
Ordinary light is made up of waves that **vibrate** in all directions, up and down, from side to side, and diagonally. A polarising filter lets through only light waves that are vibrating in a certain direction. It stops all the others. Sunglasses with polarising filters cut out bright glare.

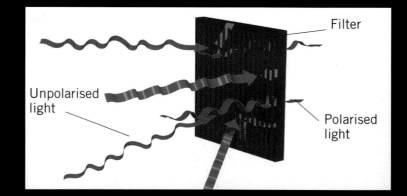

Filter

Unpolarised light

Polarised light

This polarising filter lets through only light waves that are vibrating in an up and down direction.

A lighthouse contains very bright arc lamps. The lamps make light by making a continuous spark that jumps between two electrical contacts.

HISTORY FILE

LIGHTING THE WAY

• *c.* 500,000 BC: people discovered fire and used it to make light and for cooking.

• *c.* 13,000 BC: oil lamps were invented.

• from 1,000 BC: Greeks and Romans used candles for light.

• *c.* 1800: gas lamps were used to light city streets.

• 1879: Thomas Edison in the US and Joseph Swan in England both invented the electric light bulb, with a thin filament inside. At the same time the first electricity generating stations were built so that people could use light bulbs at home.

• 1930s: the fluorescent tube (now used for energy-saving bulbs) became available to buy. It was invented in the 1890s, and produces light when electricity flows through the gas inside.

• 1980s: compact fluorescent bulbs became available. They are used as low-energy light bulbs.

MORE WAYS OF MAKING LIGHT

Some substances **absorb** light or other forms of **radiation,** and then give out a different sort of light. This is called fluorescence. For example, invisible fluorescent ink shows up when **ultraviolet** light shines on it. Other substances absorb light or other forms of radiation, too, but they release light slowly. This is called phosphorescence. Glow-in-the-dark toys use phosphorescent materials to make their glow.

ANIMAL LIGHTS

Some animals can make their own light with their bodies. They are described as bioluminescent. Some animals that live in the dark depths of the ocean, such as anglerfish and lantern fish, use light to attract prey. Others use light to attract a mate or to warn off predators.

These streaks of light come from fireflies, which are flying beetles.

Light on the Move

Light is the fastest thing in the universe. In space, where there is a vacuum, the speed of light is 300,000 kilometres per second. Scientists don't think that anything can go faster than that. Light can travel nearly that fast in the air, but it does travel slower in other substances. For example, in glass, the speed of light is about 185,000 kilometres per second. Other substances don't let light through them at all.

TRANSPARENT SUBSTANCES

Substances that light rays can travel straight through are said to be transparent. Examples of transparent substances are air, glass, water and clear plastic. We can see right through transparent substances, and see things on the other side clearly.

OPAQUE SUBSTANCES

Some substances don't let light through at all. These substances are said to be opaque. The light that hits an opaque substance either bounces off the surface or is absorbed by the object. Substances such as wood and metal are opaque.

Light always travels in straight lines, like these sunbeams.

TRANSLUCENT SUBSTANCES

Some substances let light through, but we can't see objects on the other side clearly. These substances are said to be translucent. Examples of translucent substances are tracing paper and frosted glass. They make objects look blurred because the substance scatters the light rays in different directions. Some light also bounces off translucent objects.

This layer of fog is translucent. The lights underneath are blurred.

HISTORY FILE

WAVES OR PARTICLES?
In the seventeenth century, Dutch scientist Christiaan Huygens suggested that light travels in waves. Other scientists disagreed. They said that light was made up of a stream of tiny particles. Today scientists accept both ideas. They sometimes treat light as waves and sometimes as a stream of tiny parcels of energy, called photons.

SHADOWS

Light rays always travel in straight lines. When light rays hit an opaque object they can't get round, and objects on the other side don't get any light. So the object creates a dark area, called a shadow. The very dark, central part of a shadow is called the umbra. Around it is a light area called the penumbra.

MOON AND EARTH SHADOWS

Sometimes the Moon moves between the Sun and the Earth. It blocks the sunlight, making a shadow on the Earth's surface. This is called a solar eclipse. Sometimes the situation is reversed, and the Earth moves between the Sun and the Moon. Then the Earth casts a shadow on the Moon. This is called a lunar eclipse.

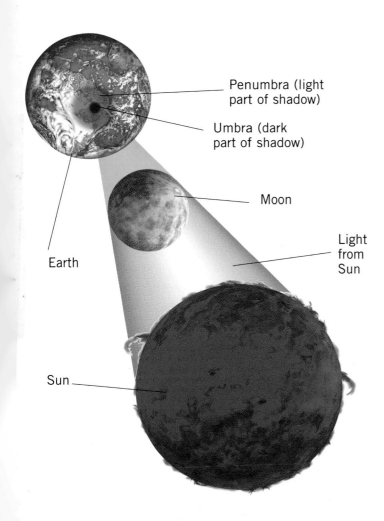

Penumbra (light part of shadow)

Umbra (dark part of shadow)

Moon

Light from Sun

Earth

Sun

TEST FILE

MAKE A SUNDIAL

• Cut a right-angled triangle from card with two short sides about 10 cm long and one side about 14 cm long. This part of the sundial is called a gnomon.

• With sticky tape attach one short side of the gnomon to the centre of a sheet of card so that the other short side stands upright.

• On a sunny day put the sundial outside. Use a compass to line up the base of the gnomon on a north-south line. The vertical edge should face north.

• At nine o'clock, ten o'clock, and so on, draw a line along the edge of the shadow made by the vertical edge of the gnomon. Write the time at the end of the line.

• On the next day you can tell the time by your sundial.

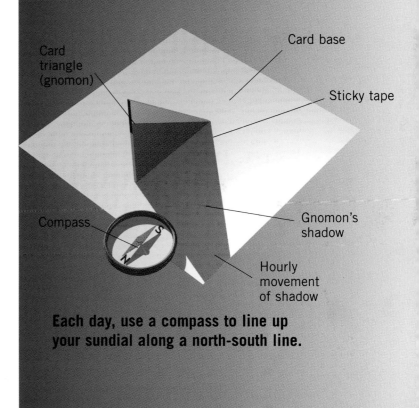

Card triangle (gnomon)

Card base

Sticky tape

Compass

Gnomon's shadow

Hourly movement of shadow

Each day, use a compass to line up your sundial along a north-south line.

A solar eclipse throws a huge shadow on to the Earth.

13

Bouncing Light

We see objects such as the Sun, light bulbs and candle flames because they are sources of light – they give out light. We see other objects, such as this book, because light from light sources bounces off them and into our eyes. The scientific word for bouncing light is **reflection**.

ROUGH AND SMOOTH SURFACES

Most objects, such as the paper in this book, have surfaces that are quite rough. Light that reflects off them is scattered randomly in all directions, and they look dull. Very smooth surfaces, such as glazed pottery, reflect light in an organised way, and they look shiny. You can see a clear reflection in them, called a **mirror image**.

When you look in a mirror, your brain thinks that the light reaching your eyes has come from behind the mirror. In fact, the light has bounced off the mirror.

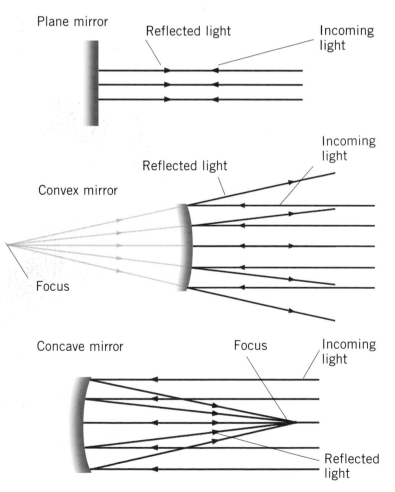

Flat, convex and concave mirrors.

FLAT AND CURVED

Most mirrors are flat (and are also called plane mirrors). They don't make things look bigger or smaller. Curved mirrors are either **convex** or **concave**. A convex mirror has a surface like the outside of a ball. It makes things look smaller than normal. A convex wing mirror on a car gives a wide view of the road behind. A concave mirror curves inwards, like the inside of a dish. It makes all the light rays from an object meet at a point called the focus. It makes things look larger than they really are. It creates a magnified image. Shaving mirrors and make-up mirrors are concave.

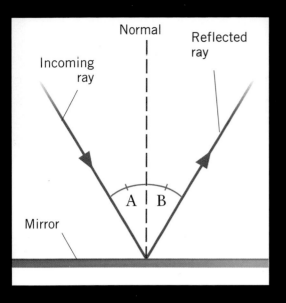

Incoming ray

Normal

Reflected ray

A | B

Mirror

INTERNAL REFLECTION

When a light ray travelling through glass hits the surface of the glass, it normally goes straight through and into the air. But if the ray hits the surface at a very low angle, it gets reflected instead and stays in the glass. This effect is called **internal reflection** because it happens inside the glass.

OPTICAL FIBRES

An optical fibre is a long, very thin piece of glass or plastic. Light travels along an optical fibre by internal reflection. When light is shone into one end, it bounces along the inside of the fibre to the other end. It even works if the fibre is bent round a corner.

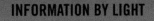
Light rays bouncing along an optical fibre.

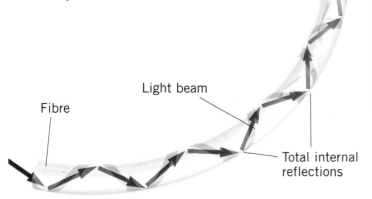

Light beam

Fibre

Total internal reflections

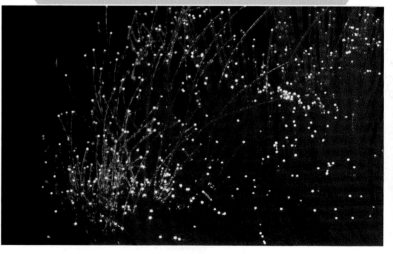

An optical fibre cable contains hundreds of optical fibres bundled together.

Bending Light

When a light ray goes out of one substance and into another, it usually bends to one side. This bending is called **refraction**. Imagine standing at the side of a swimming pool. Refraction makes the underwater swimmers look out of shape and closer to the surface than they are.

LENSES

A lens is a piece of glass or plastic with curved surfaces. Light rays are refracted as they pass through the lens. A lens is designed to bend light in an organised way. Just like curved mirrors, lenses make things appear smaller or larger than they really are. Lenses are either concave or convex.

The straw appears to be broken because refraction bends the light.

 **FACT FILE**

LAW OF REFRACTION
The law of refraction defines how a light ray bends when it moves between the air and a substance such as glass. It states that when a light ray passes into the substance, the ray bends towards a line at right angles to the surface, called the normal. When the ray leaves the substance, it bends away from the normal.

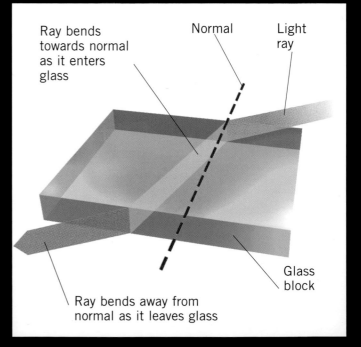

Ray bends towards normal as it enters glass

Normal

Light ray

Ray bends away from normal as it leaves glass

Glass block

A concave lens bends light rays away from each other, making things look smaller.

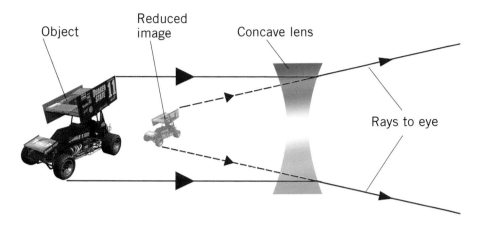

Object

Reduced image

Concave lens

Rays to eye

CONCAVE LENSES

A concave lens has surfaces that curve inwards, so that the lens is thicker at the edges than in the centre. Light rays that pass through the lens bend away from each other, or diverge. A concave lens always makes objects look smaller.

CONVEX LENSES

A convex lens has surfaces that curve outwards, so that the lens is thicker in the middle than around the edge. Light rays that pass through a convex lens bend towards each other, or converge. Two rays parallel to each other are bent so that they always meet again a certain distance from the lens. That distance is called the **focal length** of the lens.

When an object is far away from a convex lens, the lens makes an image at the focal length, which is smaller than the object. When an object is very close to a convex lens, the lens works as a magnifying glass. It makes an image larger than the object.

MIRAGES

If you are travelling along a road in hot weather, you sometimes see what appears to be a pool of water ahead. It is not water at all, but a **mirage**. A mirage appears when a layer of warm air close to the ground refracts light from the sky back upwards and into your eyes. You think you see water because the light looks as though it is reflected off the ground by water.

A convex lens used as a magnifying glass.

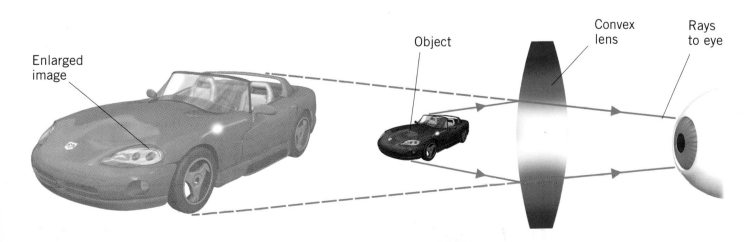

Enlarged image

Object

Convex lens

Rays to eye

Colours of Light

Light from the Sun and from light bulbs is known as white light. It looks white, but it is actually made up of many different colours of light mixed together.

You can see the colours of light if you look at objects through a piece of coloured plastic, such as a sweet wrapper. If you look through a red piece of plastic, everything on the other side looks red, and if you look through a blue piece of plastic, everything on the other side looks blue.

A piece of coloured plastic works as a filter. It absorbs some colours of light but lets others through. For example, a red filter absorbs all the colours of light except red. It lets red through to reach your eyes, so the wrapper and view look red. Similarly a blue filter absorbs all the colours except blue.

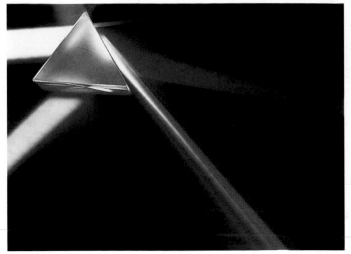

A prism splits white light into its colours.

SPLITTING UP LIGHT

We can prove that white light is made up of different colours by splitting up a beam of white light with a prism – a triangular block of glass or plastic. When light goes through the prism and out the other side, it bends, but the different colours of light bend by different amounts. Red light is bent least and violet light is bent most. The range of colours that the light splits into is called the **colour spectrum.** The main colours are red, orange, yellow, green, blue, indigo and violet.

MIXING COLOURS
Mixing pigments
• All the different colours can be made by mixing together cyan, magenta and yellow pigments in different amounts.
• Cyan, magenta and yellow are called the primary colours of pigments.
• For example, cyan paint and yellow paint mixed in equal amounts make green paint.
• All three colours mixed in equal amounts make black.

Mixing light
• All the different colours can be made by mixing together red, green and blue light in different amounts.
• Red, green and blue are called the primary colours of light.
• For example, red and green lights mixed in equal amounts make yellow light.
• All three colours mixed in equal amounts make white.

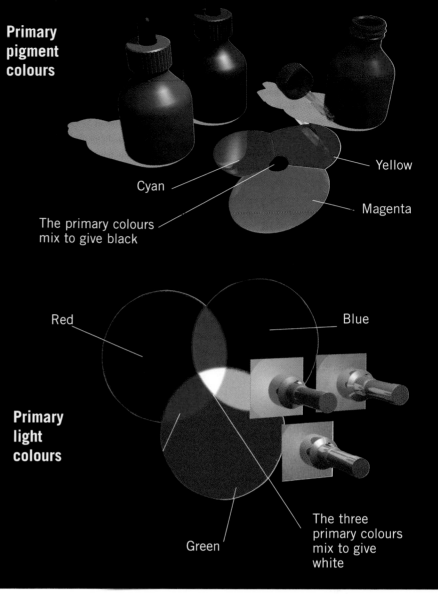

Primary pigment colours

Cyan

Yellow

Magenta

The primary colours mix to give black

Primary light colours

Red

Blue

Green

The three primary colours mix to give white

COLOURS OF THE RAINBOW

A rainbow appears when sunlight hits a shower of rain. The raindrops work like tiny prisms, splitting the sunlight into its colours. You have to stand with your back to the Sun to see a rainbow. The colours of a rainbow are the same as the colours in the spectrum, and appear in the same order.

Rainbows appear when the Sun shines through raindrops.

OBJECTS IN COLOUR

We see most objects because light bounces off them and into our eyes. These things have colour because only some of the colours of light bounce off them. All the other colours are absorbed by the objects. For example, an apple looks green because it absorbs all the colours of light except green, which bounces off. White things reflect all the colours of light, and black things absorb all the colours of light.

Objects have colour because they contain substances called pigments. Paints, dyes and inks all contain pigments.

Seeing Light

Each of your eyes is made up of an eyeball slightly smaller than a ping-pong ball. The eyeballs are soft, filled with fluid, and protected from damage by a bony eye socket.

Light gets into the eye through a dark hole called the pupil. The size of the pupil is controlled by tiny muscles that make up the coloured ring around it, called the iris. In bright light the iris makes the pupil small to stop too much light getting into the eye. In dim light it opens the pupil up to let more light in.

A fly's eye has no lens. It is made up of hundreds of tiny light-sensors.

BENDING AND FOCUSING

Light rays entering the eye from a scene are bent towards each other so they meet at the back of the eye, forming an image of the scene. The bending is done by the cornea at the very front of the eye and the lens, which is just behind the iris.

Muscles around the lens, called ciliary muscles, change the shape of the lens. They make it fatter to bend light from close-up objects into focus, and thinner to bend light from distant objects into focus.

THE RETINA

The image that the cornea and lens make is right at the back of the eye. This part of the eye is called the retina. It is covered with cells that detect light. The cells send signals to your brain along the optic nerve

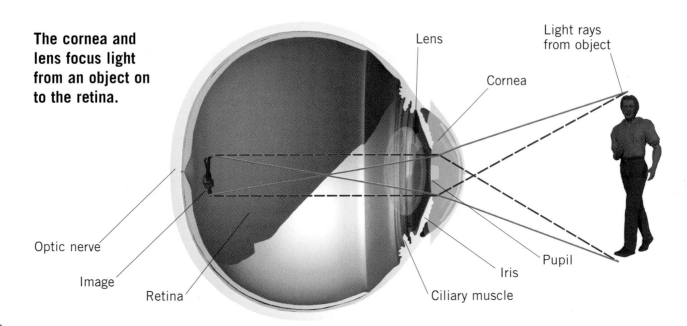

The cornea and lens focus light from an object on to the retina.

Lens

Light rays from object

Cornea

Optic nerve

Image

Retina

Ciliary muscle

Iris

Pupil

FACT FILE

JUDGING DISTANCE

Your eyes and brain work together to judge how far away objects are. Here are some of the ways that they do it.

• Hidden objects: your brain knows that objects further away can be hidden behind closer objects.

• Sizes: your brain knows that an object looks smaller the further away it is.

• Stereo vision: your two eyes get a slightly different view, which lets you see things in three-dimensions.

• Parallax: nearer objects appear to move more when you move your head from side to side.

• Accommodation: your brain knows how much it has adjusted the lenses in your eyes to focus on an object.

• Convergence: your brain knows how much your eyes are swivelled inwards to bring their two images together.

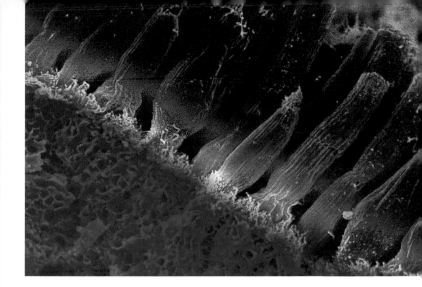

Light-detecting cells in the eye seen through a microscope.

TRICKS OF THE MIND

Sometimes your brain is tricked as it tries to make sense of the information your eyes are sending it. Something that tricks your brain like this is called an **optical illusion**. Some optical illusions are pictures drawn on paper that appear to be objects that could not exist in real life.

behind the eye. The image on the retina is upside down, but your brain turns it the right way up.

The retina has two kinds of light-sensitive cells, called rods and cones. There are about 125 million rods. They detect bright and dark, and movement. There are about seven million cones. They detect colour. The cones also let you see fine detail in the centre of the scene you are looking at.

The tribar is an object with three bars at right angles.

From a certain angle, the tribar appears to be an impossible triangle.

Optical Instruments

An optical instrument is a device that uses light to help us study small or distant things. The simplest optical instrument is the magnifying glass. This is simply a convex lens. It works by bending light from an object before it reaches your eyes. You can see how it works on page 17. It makes the light appear to come from a larger, closer object.

THE MICROSCOPE

A microscope is an optical instrument that allows scientists to see the detail on tiny objects. A simple microscope has two convex lenses in a tube. The object to be viewed, called the specimen, is placed near the bottom of the tube. The first lens, called the objective, makes a magnified image of the specimen. The second lens, called the eyepiece, magnifies that image.

FACT FILE

MICROSCOPES AND TELESCOPES

• The most powerful optical microscopes can magnify objects up to about 2,000 times.

• To magnify objects more than 2,000 times scientists use electron microscopes. An electron microscope can magnify more than a million times.

• The largest refracting telescope in the world is at the Yerkes Observatory, Wisconsin, in the USA. Its objective lens is 116 cm across. It was made in 1897.

• The world's largest telescope mirror is 9.8 metres across. It is part of the Keck Telescope in Hawaii.

A blood-sucking mite magnified 200 times by an electron microscope.

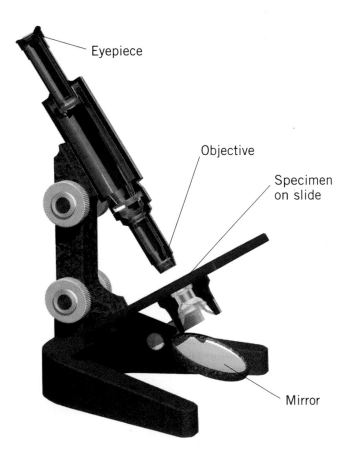

Eyepiece

Objective

Specimen on slide

Mirror

An ordinary laboratory microscope can magnify up to about 1,000 times.

THE TELESCOPE

A telescope is an optical instrument used to study distant objects, such as animals in the wild and planets in space. There are two main types of telescope – the refracting telescope and the reflecting telescope. A refracting telescope has two lenses, the objective and the eyepiece, at opposite ends of a long tube. The objective lens makes an image of the distant object. The observer looks through the eyepiece, which magnifies the image formed by the objective.

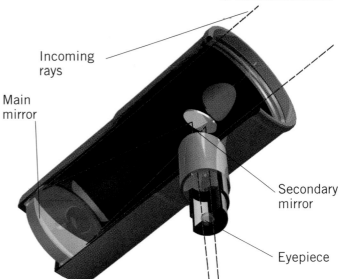

A reflecting telescope collects light with a mirror.

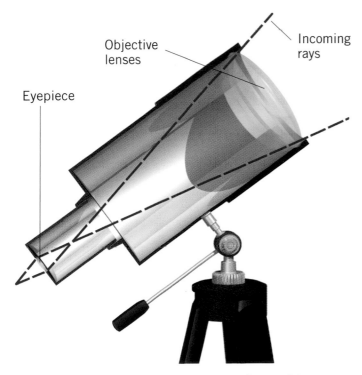

A refracting telescope collects light with a large lens.

A reflecting telescope has a concave mirror instead of the objective lens in a refracting telescope. The mirror reflects light from the object back up the tube, forming an image. The observer views the image with an eyepiece, which makes the image look larger.

TWO TELESCOPES

Binoculars are another optical instrument for studying distant objects. They are made up of a telescope for each eye, which gives a three-dimensional view. A telescope needs a long tube between the objective and the eyepiece. In binoculars the distance is made shorter by bending the light round corners using prisms in the middle of the tubes.

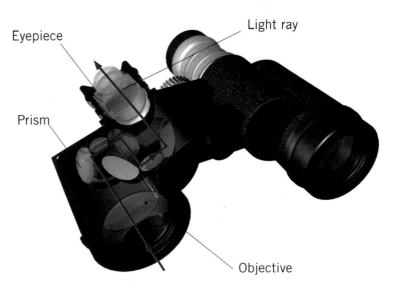

The path of light through a pair of binoculars.

Cameras

A camera is a device that records still or moving images. The two main features of a camera are a lens and a light-sensitive surface. The lens makes an image of the scene the camera is pointing at. The image falls on the light-sensitive surface, which records it.

Next to the lens is a hole called the **aperture**. This can change size to let more or less light into the camera. Between the lens and the light-sensitive surface is a shutter, which opens to let light hit the surface when a photograph is taken.

The light-sensitive surface in a camera is either a piece of photographic film or a microchip called a **charge-coupled device (CCD)**.

FILMS AND DEVELOPING

In a film camera the images are recorded on plastic film. When light hits a film, the chemicals in the film change. The film must be treated with chemicals to make the image show up. This is called developing the film.

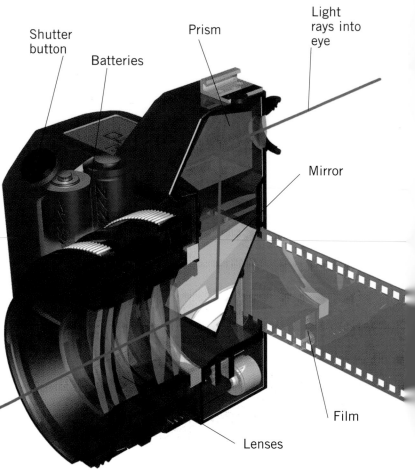

Shutter button · Batteries · Prism · Light rays into eye · Mirror · Film · Lenses

When a photograph is taken, the mirror flips up so that the light can reach the film.

HISTORY FILE

1816: Frenchman Joseph Niépce (1765-1833) made one of the first photographs. It was a negative image, with light and dark reversed, and on a sheet of metal.

1826: Niépce made the first positive photograph, showing light and dark the right way round.

1884: American inventor George Eastman (1854-1932) developed film on a plastic roll. Before this photographs were taken on metal or glass sheets.

1947: American Edwin Land (1909-91) invented the instant camera, which contained chemicals that developed the film.

1982: Sony produced the first electronic, filmless camera.

A studio TV camera takes 25 electronic pictures per second.

DIGITAL IMAGES

In a digital camera the image falls on a CCD. This is made up of thousands or millions of light-sensitive points called **pixels**. The colour and brightness of each pixel is measured electronically by the CCD. This data is recorded in the camera's memory. The memory is normally in the form of memory cards, but is sometimes a CD or DVD disc.

The greater the number of pixels on the CCD, the greater the detail that the camera can record. The number is measured in megapixels (one megapixel equals one million pixels).

The images can be transferred to a computer for viewing and editing, or sent to a printer to be printed on to paper.

MOVING PICTURES

Video cameras and television cameras record moving images. They do this by recording many still images one after the other in quick succession. Each image is called a frame, and a camera normally takes between 24 and 30 frames every second. The images are detected by a CCD and recorded on tape, normally in digital form.

When the images are shown in quick succession, on a television screen or computer screen, our brains are fooled into thinking that we are seeing a moving image.

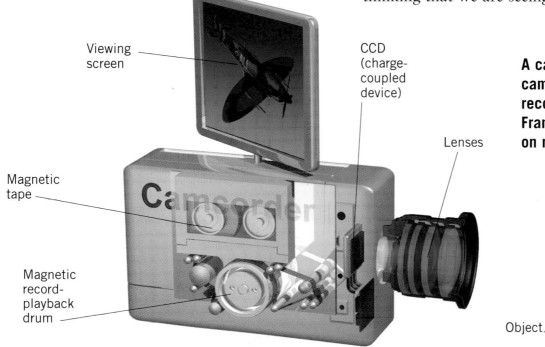

Viewing screen

CCD (charge-coupled device)

Lenses

Magnetic tape

Magnetic record-playback drum

Object

A camcorder is a camera and video recorder combined. Frames are stored on magnetic tape.

Laser Light

Ordinary light from the Sun or light bulbs is made up of many different colours of light mixed together. The waves that make up the light travel along out of step. That means their peaks and troughs do not rise and fall in step with each other. The waves of ordinary light also spread out in all directions from where they are made.

Laser light is very different to ordinary light. All the light in laser light is exactly the same colour. All the waves are in step with each other, too, which means that the peaks of all the waves are lined up. And all the waves in a beam are parallel. They all travel in exactly the same direction, without spreading out. This all means that a laser beam can carry a lot more energy than a normal light beam.

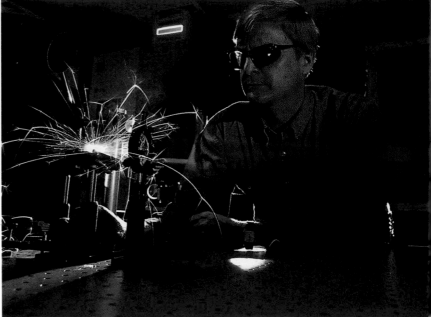

A laser beam focused by a lens, cutting through metal.

HOW A LASER WORKS

The word laser is short for 'Light Amplification by Stimulated Emission of Radiation'. It means that the intense beam of light from a laser is made by a substance that is fed energy until it gives out light.

The substance inside a laser that gives out the light is called an active medium. It can be a solid, a liquid or a gas. Different mediums give out laser light of different colours and with different energy.

Energy is fed to the medium in the form of light, heat or electricity. This makes the particles in the medium give out light. The light bounces up and down the laser tube until it is powerful enough to break out of the end, forming the laser beam.

Coiled flash tube

Laser beam

Energy-carrying particles of light

Active medium

A laser tube which is given energy by light from the coiled tube.

LASER POWER

High-powered lasers are useful in industry. When a high-power laser beam hits an object, the energy in the beam heats the object up very quickly, melting or burning a hole in it. Computer-controlled lasers are used to cut shapes from metal sheets, plastics and textiles very accurately.

Low-powered lasers have many applications too. They are used by surgeons to make small cuts during operations. These lasers are known as 'laser scalpels'. They are also used in eye surgery to shape the cornea, and so correct patients' vision. CD and DVD players contain tiny lasers that read the information on discs, and check-outs in shops use lasers to read the bar codes on their products.

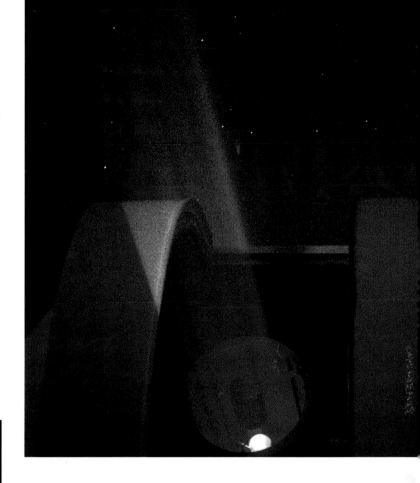

A laser beam being fired at the Moon. It can measure the distance to the Moon's surface to within 2 cm!

MEASURING WITH LASERS

Because laser beams are straight, very thin and very bright, they are useful for lining up parts of structures such as bridges and tunnels, and for checking if things are level with each other. Lasers can also be used to measure the distance between two points. A laser is fired from one point at a mirror to the other point. A device measures the time it takes for the laser beam to bounce back, and calculates the distance.

Holograms

All objects have three dimensions – height, width and depth. With our eyes we see things in three dimensions, too. That means we can see that some objects are closer than others. And if we move to one side, we can see round the side and behind objects.

Normal pictures are drawn or painted on flat paper. We say that they have two dimensions because they have only height and width. A two-dimensional picture can only show objects from one angle, and we can't see behind them by moving to one side.

A **hologram** is a picture that looks three-dimensional, but that is drawn on a flat surface. Even though a hologram is flat, it seems to have depth, as though you could reach inside it. Holograms are made with laser light.

Holograms are very hard to forge, so are used on credit cards and security cards.

INTERFERING WAVES

Holograms work because light waves interfere with each other when they meet. If the peaks and troughs of the two waves are in step with each other, they add together to make a stronger wave. But if the peaks and troughs are exactly out of step, they cancel each other out, leaving no wave at all. Where waves meet, they interfere to make patterns of light, dark and different colours.

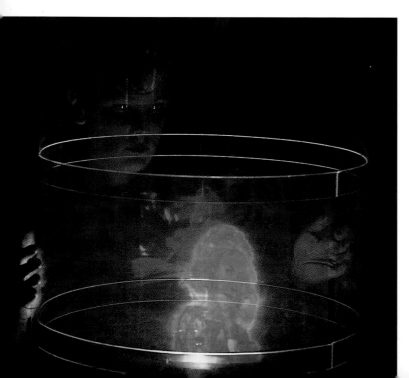

Holograms look like solid objects, but they are just patterns of light.

Interference between light waves makes these patterns on bubbles.

MAKING HOLOGRAMS

A hologram of an object is made by shining laser light on to the object. First, the laser beam is split into two beams. One beam, called the reference beam, shines straight on to a piece of special photographic film. The other beam shines on to the object. It reflects off the object and on to the film.

The film does not record a picture of the object itself. Instead, it records the interference pattern between the reference beam and the reflected object beam. This pattern is made up of tiny spots, stripes and lines, and nothing like the object at all. But when it is viewed correctly, it creates a three-dimensional view of the object.

FUTURE FILE

USING HOLOGRAMS

Holograms already have many uses, from tiny credit card logos to giant exhibition displays, and it is likely that they will have many more in the future.

- Digital, computer-generated, full-colour holograms are already possible.
- In the near future we may have cheap hologram printers linked to computers. They will make designing three-dimensional objects much easier.
- The next step could be holographic displays for computers, which could show animated holograms in real time.
- Perhaps in the end, we will have holographic television sets showing programmes and movies in three-dimensions.

Equipment for recording a hologram.

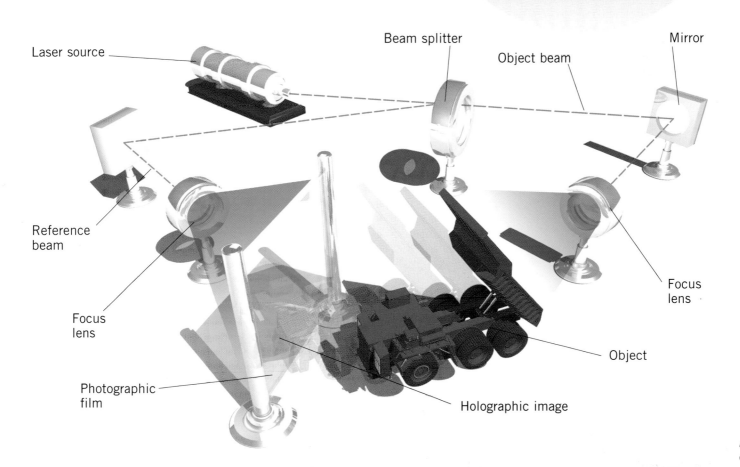

Laser source

Beam splitter

Object beam

Mirror

Reference beam

Focus lens

Focus lens

Object

Photographic film

Holographic image

What is Sound?

All the sounds we hear are made up of moving vibrations. Like light, sound is a form of energy. And like light, it moves in waves. Sounds are made when an object vibrates. For example, if you hit a drum, the skin of the drum moves up and down very quickly. This vibration makes the sound of the drum. The vibrations of the drum make the air around the drum vibrate too. The vibrations spread out in all directions through the air, just as ripples spread out on a pond. When the vibrations reach your ear, you hear the beat of the drum.

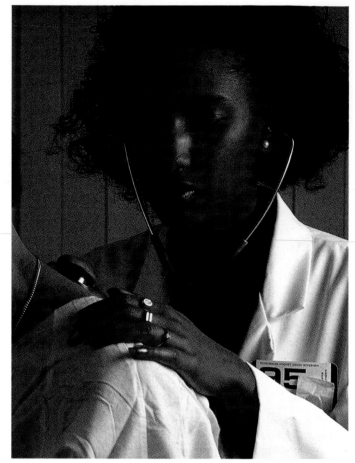

A stethoscope picks up sounds from inside your body.

PARTICLES AND PRESSURE

Sound waves are actually waves of **pressure**. As a wave passes a point, the particles at the point are squashed together then pulled apart. If you think of sound as a wave, the squashing is a peak and the stretching is a trough. The stretching creates an area of low pressure, and the squashing creates an area of high pressure. Sound waves are called longitudinal waves because the particles move backwards and forwards as a wave passes, rather than from side to side.

Sounds spreading as waves of pressure from a loudspeaker.

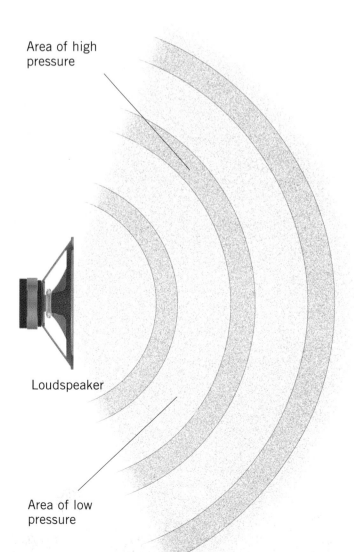

Area of high pressure

Loudspeaker

Area of low pressure

VIBRATING PARTICLES

As a sound wave moves through a substance, the particles (which are atoms or molecules) in the substance vibrate. When a particle vibrates, it makes the particle next to it vibrate too, so the sound wave moves on through the substance. The particles don't move anywhere as sound passes, they just vibrate where they are. The effect is like a cork in water, bobbing up and down as waves pass by.

A band or orchestra makes a wide range of sounds.

TEST FILE

MAKING WAVES

• Hold a 'slinky' spring stretched between your hands. Quickly push one end of the spring in and then pull it out again. A longitudinal wave moves along the spring, making the coils squash together, then stretch apart. This is like a sound wave moving through the particles of the air.

• Stretch a sheet of plastic food wrap over an empty saucepan. Sprinkle a few grains of sugar or salt on the plastic. Hold an old oven tray close to the pan and hit it hard with a wooden spoon. The sound moves through the air and makes the plastic vibrate, which shakes the grains.

Travelling Sound

Sound can travel through solids, liquids and gases, but it travels differently in each one. In a gas, such as air, the particles are not joined to each other, and they are spread apart. The particles are also swirling about because of air currents. It is difficult for sound to travel well through air because the particles often don't bump into the particles around them to pass the vibrations on. That's why sounds quickly fade away as they travel through the air.

Testing materials for their sound insulating properties.

SOUNDS IN LIQUIDS AND SOLIDS

The particles in liquids are much more tightly packed together than the particles in gases. They bump into each other much more, so sounds can travel better through liquids than gases. Sounds travel best through solids because the particles in solids are firmly attached to each other. For example, the sound of a train travels faster along the rails than through the air.

STOPPING SOUNDS

Some materials, such as plastic foam and fabrics like cotton, are not very good at transmitting sound. These materials are full of air pockets that absorb the vibrations in a wave instead of passing it on. They are useful for making sound **insulation** in houses.

FACT FILE

THE SPEED OF SOUND

When the speed of sound is written down, it normally means the speed of sound in air, at sea level. This is about 344 metres per second. If a plane flies faster than the speed of sound we say it breaks the **sound barrier**. Then the sound waves overlap each other to make a **sonic boom**.

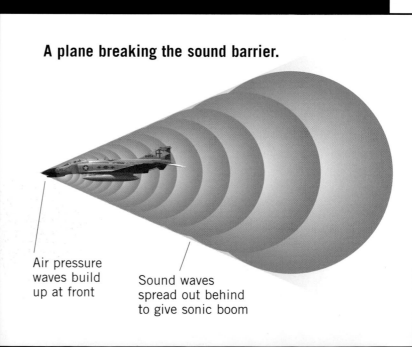

A plane breaking the sound barrier.

Air pressure waves build up at front

Sound waves spread out behind to give sonic boom

The Bell X-1, the first supersonic plane.

THE SOUND BARRIER

Today there are many planes that can break the sound barrier safely, or fly supersonically. Some can travel at more than three times the speed of sound. The first supersonic plane was the Bell X-1, which broke the sound barrier on 14 October 1947. The X-1 needed a rocket engine to make it go fast enough.

BOUNCING SOUNDS

If a sound wave moving through the air hits a solid object, such as a wall or a cliff face, it bounces back again. We say it is reflected. The reflected wave is called an echo. You can hear an echo clearly if you stand about 50 metres from a high wall or cliff and shout or clap your hands. You hear the sound when you make it and then hear it again a fraction of a second later when the echo reaches you.

SEARCHING WITH SOUND

We make use of sound to search for objects deep underwater, where it is impossible to see clearly. The sounds are made by a **sonar** machine. Sonar stands for 'SOund Navigation And Ranging'. A sonar machine sends out blips of sound into the water and listens for the echoes coming back. By sending out sounds in different directions and measuring the time the sound takes to come back, the machine can show up objects in the water. Some animals, such as dolphins and bats, use sound in a similar way to communicate, navigate and search for food.

A warship searching for a submarine using sonar. Ships use sonar to measure the depth of the water, too.

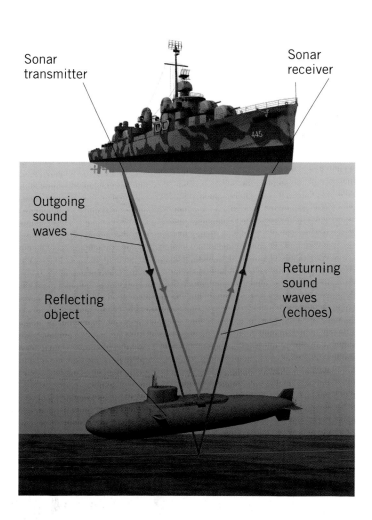

Sonar transmitter

Sonar receiver

Outgoing sound waves

Returning sound waves (echoes)

Reflecting object

Pitch and Volume

The **pitch** of a sound is how high it sounds or how low it sounds. For example, the squeak from a squeaking door has high pitch, and the rumble of a passing truck has low pitch. The pitch of a sound depends on the frequency of its waves. High-pitched sounds are made by waves with high frequency, and low-pitched sounds are made by waves of low frequency.

The **volume** of a sound is how loud the sound is. Volume depends on the amplitude of the sound waves. Waves with large amplitude sound loud and waves with low amplitude sound quiet.

MEASURING VOLUME

The volume, or loudness, of a sound is measured in units called decibels (dB). In science, decibels are used to measure the exact amount of energy in a sound wave, but decibels are a useful way to measure volume too because the volume of a sound depends on the energy in its waves. On

Jet engines make sound waves that are low in frequency and high in amplitude. They make a deep loud noise.

the decibel scale, going up by ten means the sound is ten times as loud as before.

The quietest sounds we can hear measure about 10 dB. The sound of a voice close by measures about 50 dB. Sounds that measure above 120 dB, such as a jet engine and road drill, can damage our ears. That's why airport and road workers wear ear defenders.

HEARING RANGE

We can't hear all the sounds that exist. Our ears can't detect sounds with very low pitch or very high frequency. The frequency of a sound is measured in hertz (Hz). We can't hear sounds with a frequency lower than about 20 Hz, or sounds with a frequency higher than about 20,000 Hz.

MAKE A BOTTLE XYLOPHONE
For this project you'll need a musical instrument, such as a piano or a keyboard, and somebody who knows how to play it!

• Find eight identical glass bottles, and put them in a row, each about 1 cm from the next.

• Tap the left-hand bottle with a spoon and listen to the note it makes. Find a note that matches its pitch on the piano.

• Play the next note up on the piano. Gradually pour water into the second bottle, tapping it until the note matches.

• Do the same again until all the bottles are filled with water.

• Now you can play a tune. Try this (1 is the left bottle and 8 the right bottle):
1 1 5 5 6 7 8 6 5 4 4 3 3 2 2 1

A butterfly's wings make sounds, but they are too quiet to hear.

MIXED SOUNDS

A pure sound is made up of waves of just one frequency. Pure sounds sound perfect and have perfectly shaped waves. But most sounds are not pure. They are made up of waves of different frequencies and different shaped waves mixed together. This gives them their unique sound.

For example, you can identify a person's voice because it contains a mixture of frequencies that nobody else's voice has. And you can tell the difference between a guitar and trumpet because each instrument gives out sounds with differently shaped waves.

Music in a night-club can reach more than 100 decibels.

Silent Sounds

Because we can't hear sounds with very low frequencies and very high frequencies, there are many sounds that we never hear. We can't hear sounds with frequencies lower than about 20 Hz. These are called **infrasounds.** For example, a rumbling truck makes infrasound. Even though we can't hear them, we can sometimes feel these very low sounds shaking our bodies.

USING INFRASOUNDS

Infrasounds travel much further than high-frequency sounds. The waves made by earthquakes are infrasounds that travel through the rocks of the Earth, and can be used to detect earthquakes that happen on the other side of the world. Infrasounds are sent into the ground to find oil, gas and other resources. Their echoes show up different layers of rock underground.

Some animals communicate with infrasound. For example, whales sing infrasound songs to attract mates. The sounds travel hundreds of kilometres through the ocean.

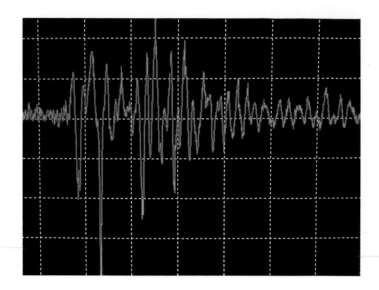

The low-frequency sound of an earthquake shown as a wave on a seismograph.

Humpback whales lie still in the water as they sing infrasonic notes.

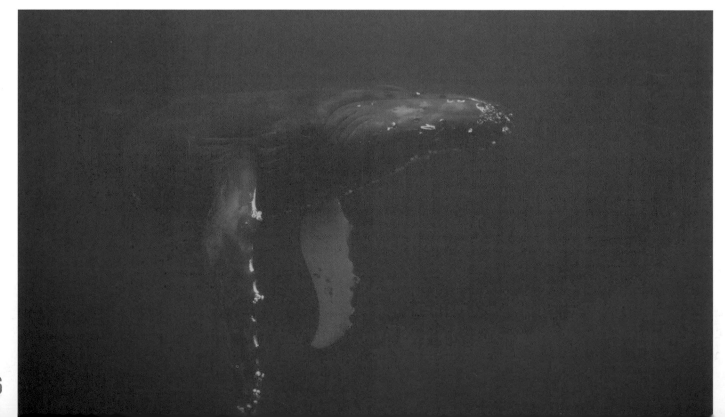

A few people can hear the ultrasounds made by bats.

ULTRASOUND

We can't hear sounds with frequencies higher than about 20,000 Hz. These are called **ultrasounds**. For example, a dog whistle makes a sound with a frequency of about 25,000 Hz. We can't hear the sound, but dogs can.

USING ULTRASOUND

We can make ultrasounds with special high-frequency loudspeakers or vibrating crystals. They have many uses in technology, science and medicine. For example, sending ultrasound through the metal of an aircraft wing can detect tiny cracks in the metal that make the metal weak. In hospitals, carefully focused beams of ultrasound are fired at kidney stones or bladder stones. The ultrasound makes the stones vibrate at high speed and shatter into small pieces. Ultrasound body scanners work like sonar machines, showing up internal parts of the body.

Animals use ultrasound, too. For example, bats send out ultrasounds and listen for echoes to see objects and prey in the dark. Grasshoppers make mating calls with frequencies up to 100,000 Hz.

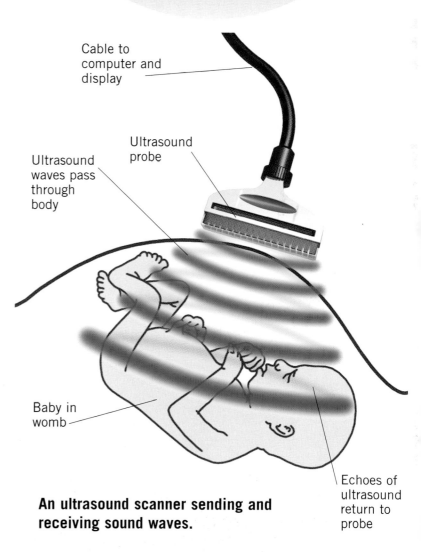

Cable to computer and display

Ultrasound probe

Ultrasound waves pass through body

Baby in womb

Echoes of ultrasound return to probe

An ultrasound scanner sending and receiving sound waves.

Hearing and Speaking

We hear sounds with our ears. The ear is a complicated organ, and the part of the ear we see on the outside of a person's head is only a part of it. The ear is divided into three parts – the outer ear, the middle ear and the inner ear. Together they collect sound and turn it into signals our brains can understand.

THE OUTER EAR

The outer ear comprises the ear flap on the outside of your head, the ear canal and the eardrum. The ear canal is the narrow channel in the centre of the outer ear. The inner end of the canal is covered with a taut piece of skin called the eardrum. The ear flap directs sound into the canal, which carries it to the drum. The sounds make the drum vibrate.

THE MIDDLE EAR

Behind the eardrum are three tiny bones connected to each other. They are called the hammer, anvil and stirrup. The hammer is connected to the inside of the eardrum. The bones carry vibrations from the eardrum to the inner ear.

Ear canal

Auditory nerve

Ear bones

Cochlea

Eardrum

Ear flap

The parts of a human ear.

THE INNER EAR

The main part of the inner ear is a coiled tube that looks like a snail, called the cochlea. It is filled with fluid, and the stirrup bone is connected to it. On the inside of the cochlea are more than a million tiny hairs. Vibrations from the stirrup make the fluid in the cochlea move about. The moving fluid makes the hairs sway from side to side. The hairs are connected to nerves, which send signals to the brain along the auditory nerve. The brain analyses the signals so that you hear the sound.

The hairs on the inside of the cochlea, seen through a microscope.

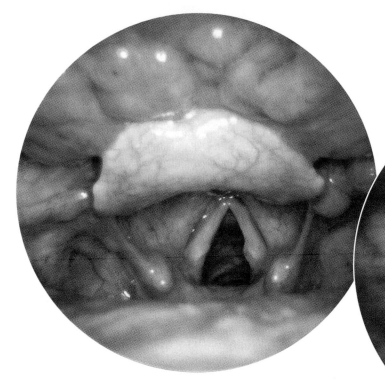

The vocal cords open for breathing and closed for speaking.

 **TEST FILE**

MODEL VOCAL CORDS
You can make a model of vocal cords with a party balloon. The neck of the balloon makes the cords, and the body of the balloon works like lungs, pushing air between the cords.

• Blow up the balloon carefully.

• Hold the sides of the very top of the neck with your thumbs and forefingers. Stretch it tight to make the hole in the neck into a narrow slit.

• Air will leak through the slit, and make the 'cords' vibrate.

• If you stretch the neck tightly the pitch of the sound will rise. If you relax your grip the pitch of the sound will get lower.

THE HUMAN VOICE

Most of the sounds we make when we speak or sing come from the larynx, or voice box, which is in the neck. The larynx is at the top of the windpipe, which is the tube that carries air to and from your lungs. The larynx contains two vocal cords, which are made of gristle. The cords are not strings, but flaps that stick out from the sides of the larynx.

MAKING SOUNDS

When you are breathing in and out, without speaking, your vocal cords are drawn back so that air can get past easily. They make a V-shaped hole. When you speak, the cords move together so that there is only a narrow slit between them. Now when you breathe out, the air flowing between the cords makes the cords vibrate. The tongue and lips also help to make sounds, such as 's' and 't'.

Acoustics and Electrical Sounds

All musical instruments work by making the air vibrate. For example, when a guitar string is plucked, it vibrates, which makes the air around it vibrate to make sound. A flute contains a tube full of air that vibrates when air is blown across a hole in its side. Instruments that make the air vibrate like this are called acoustic instruments. Electrical instruments use loudspeakers to make the air vibrate instead. Acoustic instruments can be made to sound louder, or amplified, using electricity.

Instruments can play notes with a range of frequencies and volumes. Musicians use these notes to play a series of notes and combinations of notes that are pleasing to listen to.

FACT FILE

THE DOPPLER EFFECT

Things that make sound (which are called sound sources) are sometimes moving towards or away from the person who hears the sound. This makes the pitch of the sound higher or lower. This effect is called the Doppler effect, after the Austrian physicist Christian Doppler (1803-53), who was the first scientist to describe it.

- If the sound source is moving towards the listener the waves of sound become bunched together, which makes the pitch of the sound higher.

- If the sound source is moving away from the listener the waves of sound become spread out, which makes the pitch of the sound lower.

- You can hear the Doppler effect if an emergency vehicle speeds past you. Even though the siren's frequency is the same all the time, it sounds higher as it approaches and sounds lower as it moves away.

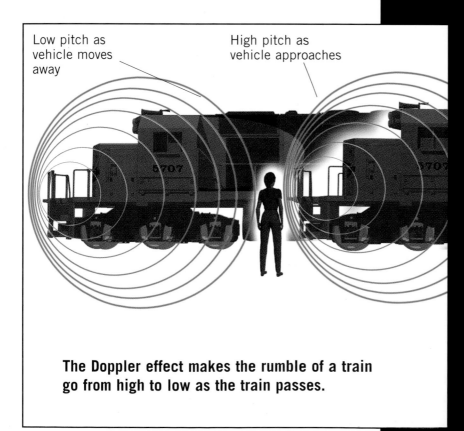

Low pitch as vehicle moves away

High pitch as vehicle approaches

The Doppler effect makes the rumble of a train go from high to low as the train passes.

SOUND TO ELECTRICITY

If we want to amplify sound (for example, in a public address system) or record a sound, we have to make a copy of it by using electricity. This is the job of a microphone. Inside a microphone is a very thin sheet of material called a diaphragm. The diaphragm vibrates up and down when sound hits it. The microphone detects this vibration and uses it to make an electric current. The changing strength and direction of the current is like a copy of the waves of the sound. This changing current is called a signal.

REPRODUCING SOUNDS

To listen to sounds that have been turned into electrical signals, we have to turn them back into sound. This is the job of a loudspeaker. Inside a loudspeaker is a coil of wire that sits inside a magnet. The coil is attached to a cardboard cone. When an electrical signal goes through the coil, the coil vibrates in and out, which makes the cone vibrate and makes sound. Signals are normally made stronger, or amplified, to make the sound from the speaker louder.

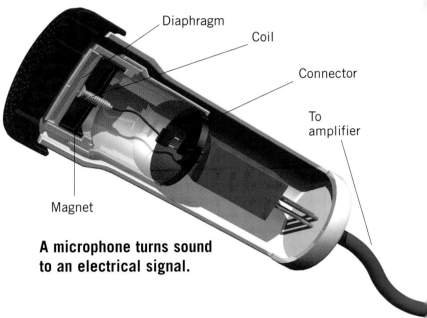

Diaphragm
Coil
Connector
To amplifier
Magnet

A microphone turns sound to an electrical signal.

A loudspeaker turns an electrical signal to sound.

Loudspeaker for high sounds

Loudspeaker for low sounds

Magnet

Coil

Cardboard cone

TEST FILE

RESONANCE
Musical instruments make full, rich sounds because the sound bounces around inside the instrument, making echoes. This is called **resonance**.

• Stretch an elastic band between your hands and pluck it. You will hear a note.

• Now find a cardboard box with a lid. Cut a round hole about 5 cm across in the lid. Stretch the elastic band across the hole and pluck it again.

• You will hear a fuller, richer note because the sound resonates in the box.

Using Sound Signals

Once a sound has been turned into an electrical signal by a microphone, it can be stored so that the sound can be reproduced later. Sound signals can also be changed by electronic circuits. They can be made larger or smaller, added together, and their shape can be changed to create all sorts of interesting effects. For example, an echo machine adds an echo to a sound.

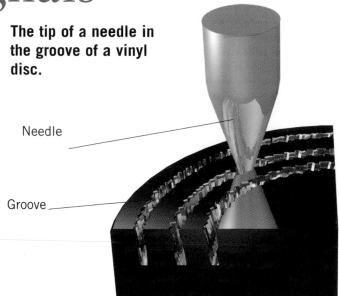

The tip of a needle in the groove of a vinyl disc.

Needle

Groove

STORING SIGNALS

We can't store an electrical signal, so we have to turn it into something we can store. A tape is covered with magnetic material. A signal is recorded on it by making tiny patches of magnetism in the material. During playback the patches are turned back into signals. On a vinyl disc the signal is used to cut a wavy groove as the disc spins. To reproduce the sound a needle is placed in the groove, which the wavy groove pushes up and down as the disc spins. This vibration is turned into an electrical signal.

DIGITAL SIGNALS

Signals are stored in computer memory and on CDs as digital signals. The original, or analogue, signal is turned into a list of binary numbers, made up of the digits 0 and 1. This is called digitisation. On a CD the numbers are stored as tiny pits and flats in the surface of a plastic disc, which are read by a laser beam inside a CD player to get the signal back.

A CD player (left) reads the microscopic pattern of pits and flats on a CD (below).

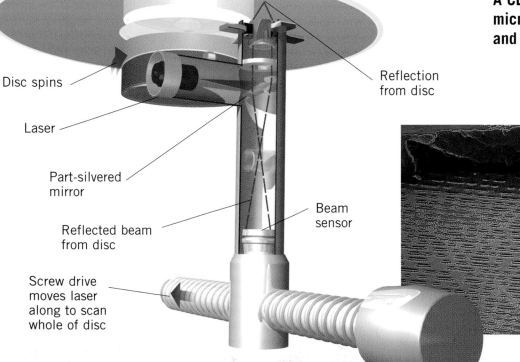

Disc spins

Laser

Part-silvered mirror

Reflected beam from disc

Reflection from disc

Beam sensor

Screw drive moves laser along to scan whole of disc

You can see the cables leading from these electric guitars to the amplifiers.

ELECTRIC INSTRUMENTS

Acoustic instruments (see page 40) make their own sound. Electric instruments, such as electric guitars, produce an electrical signal that is passed to an amplifier and loudspeaker. A device called a pick-up detects the vibrations of the strings and turns them into the electrical signal.

SYNTHETIC SOUND

A synthesiser is a musical instrument that creates electrical signals that represent sounds completely electronically. It has no moving parts like acoustic or electric instruments. The signals are sent to a loudspeaker so they can be heard. A synthesiser can produce sounds that are almost exactly the same as the sounds from dozens of different instruments, from drums to flutes.

In a recording studio sounds are represented by electricity to be mixed together.

FUTURE FILE

BIONIC SOUNDS AND SIGHTS

Sophisticated electronic circuits may soon be used in the science of bionics to help people who have impaired hearing, sight and speech.

- A microchip containing a charge-coupled device (CCD) (see page 24) could be implanted into a damaged eye. The microchip could send signals to the brain just as the retina and optic nerve do.

- A microchip in the ear could detect sound and output an electrical signal to the brain, taking the place of a damaged middle ear or cochlea.

- A microchip and artificial voice box could help people with a damaged voice box to speak clearly again. The chip could receive signals from the brain to make it operate.

Glossary

Absorb To take in and keep hold of something. For example, a sponge absorbs water.

Amplitude The height or strength of a wave.

Aperture A hole that lets light into a camera.

Charge-coupled device (CCD) A microchip that is sensitive to light, used in digital cameras.

Colour spectrum The range of colours of light that we can see. They are red, orange, yellow, green, blue, indigo and violet.

Concave Describes a lens or mirror with surfaces that curve inwards.

Convex Describes a lens or mirror with surfaces that curve outwards.

Displace To move something to one side.

Electromagnetic radiation Any form of radiation that is made up of electromagnetic waves, such as light, radio waves, microwaves and X-rays.

Focal length The distance between a lens or mirror and the point where parallel light rays bent by it meet.

Frequency The number of times something happens every second.

Hologram A flat picture that looks three-dimensional.

Infrasounds Sounds with frequencies too low for us to hear.

Insulation Material that absorbs sound, stopping it travelling to where we don't want it.

Interference When waves combine to either become stronger or cancel each other out.

Internal reflection When a light ray travelling inside glass or water bounces off the inside surface of the glass or water instead of passing through into the air.

Laser beam A narrow beam of high-energy light.

Mirage When an object appears that is not really there. It is caused by light being bent by warm air near the ground.

Mirror image The back-to-front image that appears in a mirror.

Nuclear reactions Changes that happen in the nucleus (central part) of an atom.

Optical illusion Something that fools your brain into seeing something different from what you are actually looking at.

Peak The crest, or very top, of a wave.

Pitch How low or high a sound is.

Pixels Tiny squares of colour that make up a digital picture.

Pressure The push made by a gas or liquid on objects immersed in it.

Radiation Any waves or particles of electromagnetic waves, such as light, radio waves, microwaves and X-rays.

Reflection When a wave or ray bounces off an object.

Refraction When a wave or ray bends as it crosses the boundary from one substance to another.

Resonance When the movement causing a vibration is in time with the vibration, making the vibration larger and larger.

Sonar Stands for 'SOund Navigation And Ranging'. A device that uses sound to find objects underwater.

Sonic boom A strong shock wave in the air caused by an object moving faster than the speed of sound.

Sound barrier The name given to the resistance that aircraft feel as they begin to fly faster than sound.

Trough The bottom of the dip in a wave.

Ultrasounds Sounds with frequencies too high for us to hear.

Ultraviolet A type of electromagnetic radiation that comes from the Sun, next to violet light in the electromagnetic spectrum.

Vacuum A space where there is nothing, not even air or specks of dust.

Vibrate To move one way then the other.

Volume The loudness of a sound.

Wavelength The distance between one crest and the next in a wave.

Further Information

PLACES TO VISIT

Eureka! The Museum for Children
Over 400 hands-on science exhibits, interactive activities and challenges.
Eureka! The Museum for Children, Discovery Road, Halifax, HX1 2NE
www.eureka.org.uk

The Science Museum
Thousands of exhibits and hands-on activities on science and technology.
Science Museum, Exhibition Road, London, SW7 2DD
www.sciencemuseum.org.uk

BOOKS TO READ

Discover Science: Light by Kim Taylor (Chrysalis, 2003)

Light, Sound and Electricity (The Usborne Internet-linked Library of Science) by K. Rogers, P. Clarke and A. Smith (Usborne, 2001)

Science Answers: Light: From Sun to Bulbs by Chris Cooper (Heinemann, 2004)

Science Answers: Sound: From Whisper to Rock Band by Chris Cooper (Heinemann, 2004)

Science World: Light and Lasers by Kathryn Whyman (Franklin Watts, 2003)

Science World: Sound and Music by Sally Nankivell-Aston and Dot Jackson (Franklin Watts, 2003)

WEBSITES

http://www.exploratorium.edu/explore/online.html
Examples of optical illusions

http://www.mreclipse.com
Information and photographs on solar and lunar eclipses

http://www.opticalres.com/kidoptx_f.html
Lots of information about optics for children

http://electronics.howstuffworks.com/speaker.htm
A good introduction to sound and how speakers work

Index

The Electromagnetic Spectrum

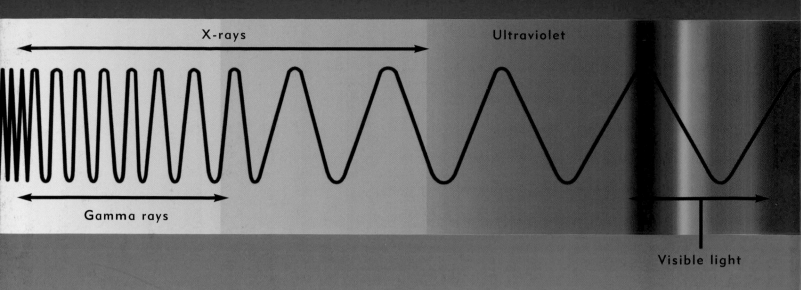

X-rays

Ultraviolet

Gamma rays

Visible light

SPEED OF LIGHT AND REFRACTIVE INDEX

Material	Speed	Refractive Index (how much the material refracts light)
Air	300.000 km s^{-1}	1.00
Water	225,000 km s^{-1}	1.33
Perspex	210,000 km s^{-1}	1.40
Glass (variable)	185,000 km s^{-1}	1.60
Diamond	125,000 km s^{-1}	2.40

Decibel Scale

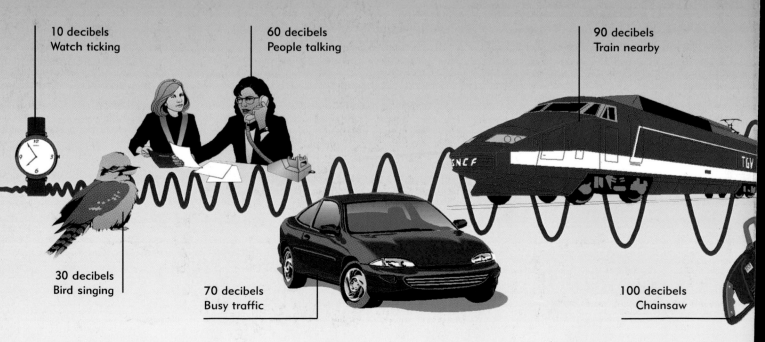

10 decibels
Watch ticking

60 decibels
People talking

90 decibels
Train nearby

30 decibels
Bird singing

70 decibels
Busy traffic

100 decibels
Chainsaw